HEAVY FREIGHT LOCOMOTIVES

GEORGE WOODS

First published 2022

Amberley Publishing
The Hill, Stroud
Gloucestershire, GL5 4EP

www.amberley-books.com

Copyright © George Woods, 2022

The right of George Woods to be identified as the Author of this work has been asserted in accordance with the Copyrights, Designs and Patents Act 1988.

ISBN 978 1 3981 0199 9 (print)
ISBN 978 1 3981 0200 2 (ebook)

All rights reserved. No part of this book may be reprinted or reproduced or utilised in any form or by any electronic, mechanical or other means, now known or hereafter invented, including photocopying and recording, or in any information storage or retrieval system, without the permission in writing from the Publishers.

British Library Cataloguing in Publication Data.
A catalogue record for this book is available from the British Library.

Origination by Amberley Publishing.
Printed in the UK.

Introduction

One of the main aims of the 1955 British Railways Modernisation Plan was the eventual elimination of steam traction during the 1970s, and its replacement by diesel and electric locos.

The first orders for diesels were placed with UK companies, some of which had very little experience in building diesels, so it was decided to build about a dozen prototypes in small numbers and, after trials, to choose the best of these for mass production.

By 1960 the railway's financial problems were causing great concern, and the government of the day encouraged BR to introduce diesels as soon as possible in an effort to cure the railway's problems. Unfortunately this was not the whole answer as both passenger and freight traffic were being lost to the roads, and the situation continued to worsen.

As part of this effort BR introduced diesel locos as fast as they could be built, without trialing them in everyday service. This resulted in some types proving to be so unreliable they were in traffic for less than ten years, but some good reliable locos were put into service – mostly those built by English Electric and Brush. At first there was not a specific freight loco, but some such as the EE classes 37, 40 and 47 were used for the heavier trains, and the less powerful types were often used in multiples.

In 1961 Dr Richard Beeching was appointed to the BTC and later became the first chairman of the British Railways board. During 1963 he produced the infamous Beeching report, which laid out the future of BR and proposed to close unprofitable lines and services in an effort to improve the efficiency of the remainder. A major part of the Beeching plan was the improvement in freight services, and one of the big changes was the switch to carrying freight in containers, which paved the way for the Freightliner service. The use of containers that could be carried by sea, road or rail, now known as intermodal, was started by the LMS in the 1930s with small 10-ton-capacity boxes, but the 1960s version was to be much more versatile, and included different types and sizes including tanks for liquids.

The Freightliner service was introduced in 1965, and ran between depots all over the BR system, where the containers could be loaded onto flat wagons and offer a fast rail service over the longer part of the transit with road haulage at each end direct to and from the customer.

Another innovation, which would revolutionise the movement of coal, was the merry-go-round service introduced in 1966 that saw dedicated coal trains running between the bigger mines and the new large power stations then coming into use. These trains were loaded and unloaded automatically at each end without the need to stop, and were usually made up of about thirty-five wagons carrying some 1,200 tons of coal. Coal destined for the domestic market was to be transported in larger air-braked wagons, and a number of coal concentration depots strategically placed around the country would handle the distribution to the customer.

At first, classes 45 and 47 locomotives were used on the Freightliner trains and classes 37 and 47 on the MGRs, but by the early 1970s it became apparent that a more powerful loco was needed. In 1976, the first Type Five freight loco, the Class 56 fitted with a Ruston Paxman 3,500-hp engine, came in to service. The first thirty were built in Romania by Electroputere, but were of such poor build quality they needed much remedial work by BR to bring them up to scratch. By 1984 another 105 were built at Doncaster and Crewe works, and were dedicated to the haulage of the MGR trains.

By the late 1970s the need for a loco that was cheaper to build and maintain than other types then in service resulted in the development of the Class 58. These were designed to use the modular system of construction, known as hood units in America, in which a substantial mainframe was used to support the engine, alternator, cabs, etc. This would be assembled in such a way that the separate parts could be replaced with refurbished components for easy maintenance. They were equipped with a 3,300-hp Ruston Paxman engine, and fifty were built at Doncaster between 1983 and 1987. In service they proved to be more reliable than the Class 56, but suffered from wheelslip problems. Unfortunately they were introduced just as the miners' strike was starting, but the reduction in coal traffic allowed them to be used on all types of freight services.

One of the most successful freight flows is the stone traffic from the Somerset quarries owned by Foster Yeoman and Amey Roadstone Construction, who send large amounts of aggregates used in the construction industry to various destinations in the south-east. Unfortunately, the BR locos used on these services were proving to be unreliable, and were causing delays and cancellation of trains. These problems led to Foster Yeoman ordering four Class 59 locos from EMD in the USA, which were delivered in 1986. Their arrival revolutionised the freight loco market in the UK. ARC then ordered another four locos that arrived in 1990. The American locos set records for reliability and load haulage that surpassed anything achieved by the BR types, typically hauling 4,300-ton trains – more than twice the load managed by the BR Class 56. Meanwhile, continuing problems with the Class 56 resulted in BR introducing the Class 60, powered by a Mirrlees Blackstone 3,100-hp engine, of which 100 were built by Brush Traction at Loughborough. After overcoming some teething problems, the first entered service in 1990 and all 100 by 1993.

As a result of the privatisation of BR in 1994, most of the freight services were taken over by EWS, which was controlled by the American-owned Winsconsin Central Railroad. They soon came to the opinion that the fleet of locos they took over was not up to the job, and set out to get rid of the worst types and attempt to improve the rest. The first types to go were the classes 20, 31, 33, 47 and 73. Some of these went to other operators, and it was decided that some Class 08, 37, 56, 58 locos, and all 100 Class 60s, would be kept.

It was decided to import 250 EMD Class 66 3,300-hp locos built by EMD in Canada at the London Ontario facility, which were an updated version of the Class 59. These came into service from 1998, and replaced the older and less reliable ex-BR types as they suffered serious failures or became due for major overhaul. Where EWS started, the other major Railfreight company, Freightliner, soon followed by also ordering Class 66 locos, eventually putting some 130 into service. In all, more than 500 Class 66s were ordered by the five major UK Railfreight operators over an eighteen-year period. Many more were ordered by various European railways.

Another American company, General Electric, tried to enter the European market with the Class 70, a 3,300-hp loco built after the style of the Class 58. These were not a great success, with only thirty-seven examples eventually being built, entering service with Freightliner and Colas Rail.

Since privatisation in 1994, many operators who tried to enter the Railfreight market fell by the wayside or were taken over by other companies.

The amount of freight carried by rail has steadily increased since 1994 with the vast increase in intermodal traffic from the major ports of Felixstowe, Thamesport and Southampton leading the way. This has made up for the loss of coal traffic since the closure of nearly all UK mines and coal-fired power stations. Other traffics that have increased include construction materials, timber, food, vehicles and waste, but since the 1980s the mixed freight train, which was such a familiar sight on all parts of the rail network since railways began, has disappeared as it largely transferred to the road. For the rail enthusiast, the variety of loco types and freight traffic has improved, with older types of locos being retained by the smaller operators and new types being introduced. This has revived the interest in freight trains. It seems that every few months a new operator comes on the scene, usually refurbishing Class 56 or Class 60 locos and entering the market, often with a new flow of traffic.

The earliest of the pictures in this book date back to the 1970s but, some fifty years later, some of these locos are still in everyday use – far outliving their planned lifespan, and recently fewer new locos have entered service. This may well be because of the uncertainty caused by climate change, as few operators want to spend large amounts on new diesel locos when in a few years legislation could ban their use. What will replace the current fleet of diesel locos is at best uncertain at the moment, but will be very important as railborne freight traffic continues to form an essential part of the economy.

Abbreviations
BG British Gypsum
BR British Railways
BSC British Steel Corporation
BTC British Transport Commission
DRS Direct Rail Services
ECML East Coast Main Line
EMD Electro Motive Diesel (Canada/USA)
EWS English Welsh & Scottish Railways
GBRF Great Britain Railfreight
HST High Speed Train (InterCity 125)
MGR Merry-go-round train
NP National Power
NRM National Railway Museum
S&C Settle & Carlisle Line
TMD Traction Maintenance Depot
WCML West Coast Main Line
VTWC Virgin Trains West Coast

Brand new No. 56001 passes through Rotherham Masborough station on a trial run from Doncaster Works on 8 June 1977. Masborough station closed on 3 August 1988 after the better situated Rotherham Central station opened on 11 May 1987.

Recently built No. 56077 is seen at the Rainhill Cavalcade paired with the advanced passenger train on 25 August 1980.

Heavy Freight Locomotives

No. 56049 waits in the murk for the road at Peterborough before heading north with fly ash empties from Fletton, heading for one of the South Yorkshire power stations in February 1979.

Also on the same day in February 1979, No. 56024 stands alongside Doncaster station with a train of MGR hoppers for one of the nearby collieries.

No. 56034 crosses from the Sheffield line to the Leeds line at Chaloners Whin Junction with northbound MGR hoppers in June 1980. Running in from the left of the picture is the ECML from Selby, which was closed along with the junction when the Selby diversion line opened in 1983.

In this busy scene at York Holgate, No. 56034 is seen heading south with a train of MGR hoppers as an HST runs into York station, and a Class 31 is about to take the station-avoiding line at Holgate Junction with a steel train for Teeside. August 1980.

Nos 56029 and 55013 spend a quiet Sunday afternoon in the yard at York TMD. In the background are York Minster and the spire of the York Oratory Church. June 1981.

No. 56122 in BR blue large logo livery braves the heavy rain as it passes through Kirkham Abbey with the White Rose Rambler rail tour that ran from Plymouth to Scarborough and return on 31 July 1983.

Railfreight grey-liveried No. 56063 has stopped in the loop at York Holgate for a crew change while heading north with oil tankers in January 1986.

No. 56116 passes Holgate in York with a rake of empty Cawoods coal containers in April 1988.

Heavy Freight Locomotives

Above and below: Two Class 56s in Coal Sector livery. Above, No. 56128 taking the York station-avoiding line in May 1990 with a train of loaded Cawoods coal containers from Blyth to Ellsmere Port. In the left-hand side of the background, coaches stand outside the York Carriage Works that closed in 1996. Below, No. 56088 stands in the yard at York TMD in June 1990.

No. 56011 in Railfreight Red Stripe livery stand alongside No. 56021 in this scene at Worksop in January 1991.

No. 56107, in Railfreight Red Stripe colours, stands in the yard at Knottingley TMD. January 1991.

Knottingley TMD was responsible for the day to day maintenance of locos working the heavy coal traffic between the Selby Coalfield and the power stations in the area east of Leeds. The depot is seen here crowded with locos on a Sunday in January 1991, with classes 08 and 56 visible. The Ferrybridge power station, which used vast amounts of coal, is very visible in the background. Of the three original power houses, only one remains in use today – but now burns waste products and biomass.

No. 56030 *Eggborough Power Station*, seen here at Berwick-upon-Tweed station in August 1992. Far from its home base of Toton, it has probably failed and is waiting a tow back home for repairs.

No. 56131 *Ellington Colliery* passes the site of Monk Fryston station, which closed in 1959, on a train of MGR hoppers in April 1997.

Above and below: Two pictures of empty steel wagon trains heading north through Colton Junction in May 1998 on their way to the Teeside steel works. The top image shows No. 56067 followed by No. 56060, both in the then new EWS colours.

Above and below: Two pictures taken in July 2000. First, No. 56058 approaches Caistor Road level crossing at New Barnetby heading for Immingham with empty MGR hoppers. About a mile up the line, with a matching field of poppies as a background, is No. 56091 Stanton passing Melton Ross with another train of empty MGRs.

Above and below: No. 56054 *British Steel Llanwern* has just left Gascogine Wood after taking on a load of coal and passes Milford Junction, followed by No. 56059 with another coal train that has come down the ECML from one of the North East collieries. Both seen on 15 February 2001.

No. 56025, wearing Transrail livery, approaches Barnetby station. with a train of imported coal travelling from Immingham docks to one of the South Yorkshire power stations in March 1998.

No. 56018 has just passed Colton Junction and heads south down the ECML with a train of HEA wagons carrying domestic coal. July 2000.

Showing off its new Load Haul colours, No. 56085 approaches Chesterfield on 28 September 2001 with a train of coal for Radcliffe power station.

No. 56054 *British Steel Llanwern* and No. 56065 stop at Warrington Bank Quay station to change crew on 4 April 2002 while heading a southbound freight.

Above and below: Two pictures taken at Doncaster station on 18 April 2002. The first shows No. 56033 *Shotton Paper Mill* passing through light engine, followed by No. 56085 heading towards Scunthorpe with a train of empty steel wagons.

Above and below: Two pictures taken at Barnetby station. First No. 56018 heads towards Scunthorpe steelworks with a load of coal, and in the other direction No. 56037 passes through with a lengthy train of assorted wagons of steel for export, heading for the docks at Immingham on 22 May 2002.

No. 56049 carries a mixture of Transrail and BR Dutch livery as it rolls downhill at Melton Ross with a mixed freight. 22 May 2002.

No. 56099, plus a dead No. 66072, pass through Carlisle Citadel station with the early afternoon civil engineers train for Crewe, which was routed via the Settle & Carlisle line. Taken on 10 June 2003.

Above and below: No. 56115 *Barry Needham* has just arrived at York station with the Twilight Grids railtour from Bristol on 31 March 2004. At the other end, No. 56078 waits to leave on a mini tour to Manchester and back. Although meant to mark the end of Class 56s in service, this rail tour proved to be premature as many Class 56s have been refurbished and remain in regular service today.

Nos 56312/67022 speed north through Appleby station with the Festive Settle & Carlisle Explorer rail tour from Derby to Carlisle on 29 December 2008.

No. 56312 is seen again in another guise, this time in a special livery to advertise the NRM Railfest at York. Taken just before the skies opened on 6 June 2012.

No. 56049 is seen at the Crewe open day on 8 June 2019 in the livery of its new owners, Colas Rail.

No. 57012 *Freightliner Envoy* passes through Stratford station in September 2000 with a container train from Felixstowe. Freightliner took delivery of twelve Class 57s, which they used from 1998 until 2008, when they were replaced by new Class 66s.

Porterbrook-liveried No. 57601 has brought in the stock to form a train for the West Country at Paddington station on 17 April 2002.

Above and below: Two shots of No. 57308 *Tin Tin* heading the Carlisle to Chirk Kronospan log train. Above, it is at Greenholme rolling down the bank from Shap to Tebay on 16 July 2007, and below it is seen waiting to head south in the middle road at Carlisle station on 19 September 2007. Thirty-three former Class 47s were rebuilt with EMD engines and sixteen were used by VTWC as Thunderbirds to haul services over non electrified routes, and to rescue failed trains. All the VTWC 57s were named after characters from the ITV *Thunderbirds* television series.

Above and below: Two diverted VTWC trains seen near Birkett Tunnel on the S&C line. First, No. 57310 *Kyrano* exits the tunnel with a diverted Euston to Glasgow train on 29 April 2006. A year later, on 26 May 2007, No. 57315 *The Mole* climbs to the tunnel with a southbound train.

No. 57307 *Lady Penelope* has just passed through Kirkby Stephen on 2 June 2007 with a Euston to Glasgow train formed of No. 390029, which is advertising the stage show *Monkey: Journey to the West*.

No. 57601, in West Coast Railways livery, stabled in the west siding at York station on 28 May 2008. Just visible above the loco is the York Eye, which was dismantled in 2013.

No. 57316 still in its Arriva Trains blue livery, although now owned by WCRC, waits in the middle road at Carlisle station before heading to the DRS depot at Kingmoor on 16 July 2013.

No. 57307 still has its *Lady Penelope* nameplates, despite now being part of the DRS fleet. It is seen here at the open day at Crewe on 8 June 2019.

No. 58011 heads south approaching Chesterfield with a coal train for Radcliffe power station in October 1990.

Above and below: On a Sunday in January 1991, Nos 58033 and 58023 are seen in the yard at Barrow Hill TMD, and later the same day No. 58048 stands on one of the loop lines at Worksop waiting to resume work on Monday morning.

Highly polished for the occasion, No. 58046 *Ashfordby Mine* and No. 58002 *Daw Mill Colliery*, both in Mainline livery, roar up the ECML passing Shipton with the Worksop Aberdonian rail tour, shortly after sunrise on 21 September 1996.

Above and below: Six months later, No. 58046 still carries the painted rail tour details as it leaves Gascoigne Wood with a coal train on 17 March 1997. Shortly after, No. 58027 follows with another load.

No. 58027 passes Monk Fryston with yet another coal train on 3 October 1997. Traffic must be thin on the ground since Gascoigne Wood loading point closed along with the Selby coalfield in 2004.

No. 58001 was repainted into its original livery for display at the Doncaster Works open day, which was held on 26 July 2003.

Above and below: Two views of No. 59001 *Yeoman Endeavour*. First in its original Yeoman livery at Sheffield Tinsley TMD open day on 29 September 1990. Then after the takeover by Aggregate Industries in 2006, No. 59001 is seen in their livery at Eastleigh Works open day on 25 May 2009.

No. 59001 passes through Reading station with a train of empties for the quarry on 24 August 2017.

No. 59103 *Village of Mells* attended the Eastleigh Works open day on 25 May 2009.

Above and below: Two views of National Power Class 59/2s near the Gascoigne Wood loading point. On 17 March 1997, No. 59202 *Vale of White Horse* arrives with empties and No. 59204 *Vale of Glamorgan* departs with a loaded coal train on 3 October 1997.

No. 59201 *Vale of York* passes Monk Fryston with a coal train on 3 October 1997.

No. 59206 *Pride of Ferrybridge* waits for the road at Stratford station on 4 June 2003.

No. 59206 *John F Yeoman* is one of the exhibits at the Eastleigh Works open day on 25 May 2009.

Above and below: Two views of No. 60095 *Crib Goch* in Railfreight Construction livery, working ordinary passenger trains from Preston to Barrow and back for Regional Railways on 25 April 1992. It is first seen crossing Levens Viaduct with a train for Barrow, and then seen leaving Grange Over Sands station for Preston.

No. 60071 *Dorothy Garrod* passes Grange Over Sands station with a sand train in June 1992.

No. 60049 *Scarfell* passes Thrimby Grange with empty stone wagons for Shap Quarry on 30 July 1994.

Above and below: No. 60061 *Alexander Graham Bell* in Transrail livery. Above, it has just passed through Blea Moor tunnel on 6 February 1999 as it makes its way south with a coal train. Below, it has just passed Colton Junction as it heads south down the ECML with a tanker train in July 2000.

Above and below: Two pictures of No. 60002 *Capability Brown*. The first picture shows it in Railfreight Petroleum livery shortly after leaving Birkett tunnel with a train for the British Gypsum plant at Kirkby Thore on 14 July 1996, but by 3 May 1997 it had been repainted into EWS colours and is seen descending from Shap at Greenholme with a train of tankers.

No. 60083 *Shining Tor* smokes it up. It has just got the road into the loop at Low Fell while heading a southbound coal train to await the passing of an express on 3 May 1997.

Above and below: Two more pictures on 3 May 1997 at Low Fell. The first is of No. 60022 passing through with a train of empty Cawoods coal containers, and the second shows No. 60044 *Alisa Craig* in Mainline blue with a train of tankers. This was a great day for Class 60s – four in two locations in about three hours was unusual, even back then.

No. 60008 *Gypsum Queen II*, in Load Haul livery, crosses from the Sheffield line at Colton Junction with a train of steel for Teeside in August 1998.

Nos 60035–60039 have just crossed Ribblehead Viaduct and are approaching Blea Moor with gypsum containers for the British Gypsum plant at Kirby Thore on 23 March 2002.

Above and below: Two pictures taken at Melton Ross on 22 May 2002. No. 60038 *Bidean Nam Bian* passes with tankers for Immingham, then No. 60062 in Transrail livery goes in the other direction with iron ore for the BSC Works at Scunthorpe.

Three pictures of the Drax to Kirkby Thore gypsum trains. First, No. 60030 *Cir Mhor*, passing over the 1,163-foot summit at Ais Gill on 14 May 2003. Also seen is No. 60036 rolling downhill through Waitby on 2 August 2003, and finally No. 60024 heading south through Appleby station on 8 March 2007.

No. 60052 *Glofa Twr-Tower Colliery* passes through Stratford station with a train of hoppers on 4 June 2003. The surroundings at Stratford have changed completely since this picture was taken, with the construction of the Olympic Stadium and park, along with the rebuilding of the station, have transformed the area.

No. 60017 cruises past grazing sheep as it approaches Kirkby Stephen on 2 June 2007. The view here remains the same, only the sheep have changed.

Heavy Freight Locomotives 49

No. 60023 braves the rain as it passes through Barnetby with a load of iron ore from Immingham Docks to the BSC works at Scunthorpe on 27 May 2008.

No. 60078 *Teenage Spirit* is on display at the open day on 26 July 2008 at Steamtown Carnforth, which is now the main workshops and maintenance facility for WCRC.

No. 60088 passes Appleby station with a short southbound Railtrack train for Crewe on 29 December 2008.

Another light load is behind No. 60085, making for Immingham on 16 September 2009 with a pair of bogie tankers during a break in the clouds at Melton Ross.

In contrast to the two previous shots, No. 60054 has a lengthy train of tanks in tow as it passes Melton Ross on 30 October 2019.

Looking very smart in their fresh Colas livery, Nos 60002 and 60085 pass through Doncaster station on 14 February 2016.

Above and below: Two pictures taken at Monk Fryston on 15 February 2001. Above, No. 66083 passes with an MGR train from Gascoigne Wood and, judging by the sign in the cab windscreen, Bill Knot is at the controls. Below, No. 66245 is passing with a steel slab train from Teeside. Driver incognito.

Above and below: In 1995 National Power started to run its own coal trains from Gascoigne Wood to Drax power station, and purchased six Class 59/2 locos and 106 bogie wagons for this service. But in 1998 National Power sold the locos and wagons to EWS and after a demerger went out of business in 2001. These two pictures taken on 15 February 2001 show EWS No. 66195 passing Monk Fryston, and No. 66226 leaving Gascoigne Wood with trains of debranded NP wagons.

Taken from the road bridge at Monk Fryston, No. 66045 has a train of empty MGR wagons in tow on 15 February 2001. The large buildings in the centre of the picture are the Old Maltings.

No. 66049 waits alongside Warrington Bank Quay station with four car-carrying wagons on 4 April 2002.

No. 66553 passes south down the Midland Main Line through Leagrave station with a bin liner train on 19 April 2002.

No. 66193 passes through Stratford station on 4 June 2003 and takes the line towards Temple Mills marshalling yard and Cambridge, with a train of Lafarge stone hoppers.

Almost in the shadow of Wild Boar Fell, No. 66018 climbs up the final few yards to the 1,163-foot Ais Gill summit with a train of gypsum containers making for the Drax power station on 16 July 2003.

No. 66207 approaches Twyford with a stone train on 3 June 2004.

Heavy Freight Locomotives

No. 66102 passes through Carlisle station with a train of empty coal hoppers on 25 June 2004.

No. 66524 hauling a train of coal hoppers has stopped in York station to change drivers on 2 July 2004.

Above and below: Empty gypsum container train returning to Drax power station in the evening. Above, GBRF No. 66708 passing through Appleby station on 27 July 2005. Below, No. 66705 *Golden Jubilee* climbs towards Birkett tunnel on 29 April 2006.

Above and below: The china clay slurry working from Burngullow in Cornwall to Irvine in Ayrshire, diverted from the WCML to the S&C because of engineering works. In the above image, No. 66168 at Lunds has just passed though Moorcock tunnel on 6 May 2006. Below, a year later on 2 June 2007, No. 66233 is running downgrade at Waitby.

Above and below: Two pictures of southbound trains passing through the delightful Appleby station. Above, No. 66151 is seen on 26 September 2006 with the midday empty gypsum containers, and below No. 66530 has a train of Freightliner coal wagons bound for one of the power stations on 8 March 2007.

Above and below: Two pictures of coal trains composed of HAA type wagons, which at this time were approaching the end of their service. The above picture shows No. 66018 passing Greenholme on the climb to Shap summit on 27 September 2006 with empties for one of the Ayrshire collieries, while below No. 66198 is approaching Shotlock Hill tunnel on the S&C with a loaded train on 30 September 2006.

Above and below: The Tesco container train runs six days a week from Daventry to Mossend, which at this time in 2007 was a Stobart Rail service. In 2011 the service was taken over by DRS, who still run it today. The above image, taken on 31 May 2007, shows No. 66411 *Eddie the Engine* passing through Carlisle station. Below, No. 66411 is seen again, this time climbing to Shap summit at Greenholme on 16 July 2007, with a train of twenty-six containers.

Above and below: Two more scenes at Appleby station. Above, No. 66583, with a train of Freightliner coal wagons, heads north on 6 September 2007. Below, two days later, GBRF No. 66723 gets a signal check from the preceding passenger train while hauling the evening gypsum empties.

Above and below: Two trains are seen passing southbound through Carlisle station on 19 September 2007. Above, No.66230 is with a Network Rail train that ran to Crewe via the S&C. Below, No. 66200 Railway Heritage works a EWS coal train.

Above and below: Freightliner coal trains are seen near Garsdale station on 28 September 2007. Above, No. 66565 with northbound empties crosses Dandry Mire Viaduct, and below No. 66564 heads south with a loaded train. The station and the adjacent cottages can be seen in the background.

No. 66721, in its Renewing the Tube livery, is a long way from home as it passes the Midland Railway signal box at Garsdale station on 28 September 2007 with a load of gypsum for Kirkby Thore.

No. 66558 passes Blea Moor with northbound coal empties on 13 March 2008. In the background the top of Ingleborough, 2,372 feet high, is covered in cloud and another shower was on the way.

Above and below: South of Blea Moor tunnel, loaded coal trains from the two major players in the movement of coal. Above is No. 66150 with an EWS train, and below No. 66559 is in charge of a Freightliner service that has stopped at the signal.

Above and below: The south end of Ribblehead station, with attractive lamps erected when the station was refurbished. Above, No. 66586 passes on a dark and wet 13 March 2008, when the extra light came in quite useful. Below, No. 66177, complete with experimental white cab roofs, enjoys a sunny interval on 5 August 2009.

Above and below: Two pictures taken at Barnetby station on 27 May 2008. Above, No. 66121 passes with a coal train from Immingham Docks to BCS Scunthorpe, then ten minutes later No. 66132 comes through in the opposite direction with a train for Immingham Docks.

Above and below: Two shots taken at Melton Ross, also on 27 May 2008. Above, Metronet No. 66722 *Sir Edwin Watkin* passes with GBRF coal empties for Immingham. Below, coming down the gradient into Barnetby, is No. 66561 with a Freightliner coal train.

Above and below: Freightliner coal trains that ran over the S&C. Above, No. 66551 passes through Appleby station with a southbound loaded train on 2 December 2008. Below, No. 66616 has stopped in Carlisle station for a crew change with northbound empties on 25 June 2009.

Above and below: Brockelsby station on 16 September 2009. Above, No. 66611 comes off the Immingham line with a tanker train, and below No. 66133 passes with an iron ore train for BSC at Scunthorpe. The lines going off to the right go to Grimsby and Cleethorpes. The station opened in 1848 and closed on 3 October 1993. Both the station building and the signal box are Grade II listed.

Looking in the other direction at Brockelsby, No. 66723 *Chinook* takes the Immingham line at the junction with GBRF coal empties on 16 September 2009.

Just to the East of Barnetby is the Singleton Birch lime works, which is being passed by No. 66059 working a coal train for BSC at Scunthorpe on 16 September 2009.

Above and below: Melton Ross on 16 September 2009. Above, No. 66144 rolls down the slow line with an iron ore train for BSC at Scunthorpe, which is about to be overtaken by a Cleethorpes to Manchester Airport train, and below No. 66046 comes in the other direction with a train of coal empties for Immingham Dock.

Above and below: Two views of the thrice-weekly Mossend to Clitheroe Castle empty cement train. Above, No. 66183 passes Appleby station on 19 January 2010. Below, with the last of the winter snow still on the top of Cross Fell, No. 66101 passes Keld on 23 April 2010.

No. 66132 waits in the siding at the British Gypsum Plant in Kirkby Thore for the last loaded container to be emptied before departing for the Drax power station for another load. The gypsum is used in the manufacture of plaster board, and at this time two train loads a day were arriving at the plant.

No. 66175 climbing past Birkett Common with the gypsum empties for Drax on 19 March 2010.

No. 66185, with coal hoppers, drifts quietly downhill past a ewe and her two lambs near Waitby on 28 April 2010.

Above and below: More Freightliner coal trains in Cumbria. Above, No. 66518 climbs to Shap Summit through Greenholme with northbound empties on 24 July 2010. Below, No. 66953 heads north through Keld with more empties on 2 March 2011.

Above and below: The Carlisle to Chirk Kronospan timber train on the S&C line being worked by Colas Class 66s. Above, No. 66842 passes Appleby station on 19 January 2011. Below, No. 66843 is about to reach the 1,163-foot summit of the line at Ais Gill on 27 June 2011.

Above and below: No. 66845 climbing through Keld just north of Appleby on 20 May 2011. The aroma of freshly cut timber must make this the sweetest smelling freight running on BR.

Above and below: Smardale, 27 May 2011. Above, the timber train for Chirk is hauled by DRS No. 66426, and below No. 66531 has charge of another load of coal as it clambers uphill, through the cutting, with its 2,000-ton load at a steady 25 mph.

Above and below: The DRS open day held at their Carlisle Kingmoor Maintenance Depot on 16 July 2011. Above, DRS No. 66420 and Colas No. 66843 stand in the sun. Below, No. 66426 is on the jacks, with Nos 47790 and 57007 in the background.

No. 66094 runs downhill past Birkett Common with the loaded Castle Cement Clitheroe to Mossend train in the beautiful evening light of 27 July 2011.

Fastline No. 66434 heads a southbound container train through Carlisle station on 16 July 2011.

The Freightliner/Shanks-liveried No. 66522 climbs past Scout Green on the ascent to Shap summit with northbound coal empties on 28 January 2012.

Above and below: Two pictures taken at the 181-yard-long Crosby Garrett tunnel on 2 March 2012. Above, No. 66850 exits the south end of the tunnel with the Carlisle to Chirk log train. Below, No. 66171 leaves the north end with the gypsum train for Kirkby Thore.

No. 66074 crosses over Castle Street on the Castlefield Viaduct as it passes through the centre of Manchester with a container train from Trafford Park International Freight Centre on 23 August 2011.

No. 66547 passes under the Midland Railway footbridge at Kirkby Stephen station with a southbound coal train on 2 March 2012. The signal cabin, seen in front of the loco, replaced the original Midland box in 1974, and the goods shed beyond dates from 1876 but is no longer in railway use.

Heavy Freight Locomotives

Above and below: Crosby Garrett on 6 March 2012. Above, No. 66561 climbs across Crosby Garrett Viaduct with a southbound coal train, and below No. 66174 passes with the afternoon Railtrack train from Carlisle to Crewe. The footbridge seen behind the rear of the train marks the position of the former station, which closed on 6 October 1952.

The afternoon Carlisle to Crewe Railtrack train is famous for the wide variety of vehicles seen on this service. Above, No. 66431 passes Birkett Common on 8 August 2013; in the middle image, No. 66431 is again in charge almost at the Ais Gill summit on 6 August 2013. Below, No. 66848 roars through Appleby station with the log train from Carlisle to Chirk on 17 July 2012. Behind the loco is the water tank and crane that were installed in 1991 to replenish the many steam specials that traverse the S&C.

Above, No. 66101 in DB Schenker livery passes through Carlisle station with the empty Castle Cement tanks from Mossend to Clitheroe on 13 June 2014. In the middle image, GBRF No. 66747 passes through Doncaster station on 8 February 2015 with a coal train heading towards Belmont Yard. Below, No. 66094 rolls slowly through Sheffield station on 7 July 2015 with Cemtex hoppers, en route from Peak Forest to Selby.

Above, DRS No. 66301 stands in the west sidings at York station awaiting its next duty on 5 August 2015. In the other siding, snow plough ZZA ADB 965206 has a much longer wait before it next sees action. In the middle image, wearing the new Freightliner livery, No. 66528 *Madge Elliot MBE* is between duties at Doncaster station on 25 February 2016. Below, No. 66137 is on the rear of the weedkiller train passing my local station at Hutton Cranswick on Sunday 3 May 2020.

Above and below: Two GBRF trains are seen passing through Doncaster station. Above, No. 66722 works a northbound WBB Minerals train, and below No. 66766 heads north on a coal train on 26 October 2016.

Above and below: Two more pictures taken at Doncaster station on 27 February 2019. Above, Nos 66419 and 66529 take a heavily loaded southbound container train through, and below No. 66749, which is usually reserved for hauling the Belmond Royal Scotsman luxury train, is slumming it on menial duties with a single tanker.

No. 66133 makes for a colourful picture heading a train of biomass wagons from Immingham Renewable Fuels Terminal to Drax power station through Melton Ross on 30 October 2019.

No. 70013 makes an impressive sight heading a southbound coal train through Appleby station on 23 January 2012.

Above and below: Two views of Freightliner Class 70 locos. Above, No. 70010 passes north through Carlisle station with coal empties on 2 August 2011. Below, No. 70008 is seen at the open day at Crewe on 8 June 2019.

Above and below: Two views of Colas Class 70 locos. Above, No. 70808 is at the Crewe open day, and below No. 70809 heads a short tanker train through Melton Ross on 30 October 2019. Having seen the success that General Motors was having in Europe with its Class 59 and Class 66 locos, the Class 70 was General Electric's effort to break into Europe. The market for diesel locos, though, seems to be drying up as railways all over Europe are looking for greener alternatives.